FROM WHEEL TO WEB
40 REMARKABLE INVENTIONS

PA BOOKS

While every precaution has been taken in the preparation of this book, the publisher assumes no responsibility for errors or omissions, or for damages resulting from the use of the information contained herein.

FROM WHEEL TO WEB: 40 REMARKABLE INVENTIONS

First edition. October 2, 2023.

ISBN: 979-8215869154

Written by PA BOOKS.

Chapter 1
The Wheel

Introduction:

The Wheel, one of humanity's earliest innovations, stands as a testament to human ingenuity. It's simple yet revolutionary design has left an indelible mark on the course of history, transforming the way we move goods and ourselves.

The Dawn of the Wheel:

The story of the wheel begins over 5,000 years ago, during the Bronze Age. Initially, the wheel was not used for transportation but served as a tool for pottery and pottery making. Its circular form, however, would soon inspire a transportation revolution.

Innovative Design:

What makes the wheel so remarkable is its elegant simplicity. A circular object with a hole in the centre, it reduces friction between a moving object and its surface. When mounted on an axle, the wheel could roll smoothly, requiring far less effort than dragging or carrying loads.

Wheel Across Civilizations:

As various ancient civilizations adopted the wheel, its applications multiplied. In Mesopotamia, wheeled vehicles transformed agriculture, making ploughing and harvesting more efficient. In Egypt, the wheel found use in creating potter's wheels and chariots, symbols of power and progress. The Indus Valley Civilization also embraced the wheel, and it played a vital role in trade networks.

Impact on Transportation:

The true breakthrough came when humans harnessed the wheel for transportation. Early wheeled vehicles, such as carts and wagons, became essential for the movement of goods and people. They expedited trade, facilitated exploration, and played pivotal roles in the construction of monumental structures like the pyramids and Stonehenge.

Legacy and Continued Evolution:

The legacy of the wheel endures to this day. Modern transportation, from bicycles to automobiles, relies on the same fundamental principle of the wheel and axle. The wheel's influence extends beyond the terrestrial realm, as spacecraft and rovers on distant planets employ wheels for mobility.

Conclusion:

The humble wheel, initially conceived for pottery, became the catalyst for a transportation revolution. Its innovation reshaped societies, economies, and cultures, enabling the movement of people and goods across vast distances.

Chapter 2
Harnessing Fire

Introduction:

Fire, one of the most fundamental discoveries in human history, holds a pivotal place in the evolution of our species. This chapter delves into the profound impact of harnessing fire, a turning point that transformed the course of human history.

The Ancient Discovery:

The discovery of fire dates back over a million years when early humans, through chance or necessity, witnessed the awesome power of this natural phenomenon. Lightning strikes and volcanic eruptions likely provided the first encounters with fire. These early humans soon realized that fire could be controlled and put to practical use.

Controlled Fire: The First Innovation:

The ability to harness fire marked a monumental innovation. By creating fire through friction or by carrying a smouldering ember, early humans unlocked a new world of possibilities. Fire provided warmth, protection from predators, and a source of light during the dark hours, which greatly extended the day's productivity.

Cooking and Nutrition:

One of the most significant impacts of controlled fire was the development of cooking. Cooking food over a controlled flame not only made it safer to eat but also made it more digestible and nutritious. This shift in diet played a critical role in human evolution, allowing for the development of larger brains and a more efficient use of energy.

Social and Cultural Evolution:

Fire also played a crucial role in the development of human social structures. Gathering around a fire encouraged communal living, fostering a sense of security and togetherness. It became a focal point for storytelling, the exchange of knowledge, and the development of early language.

Toolmaking and Expansion:

The use of fire furthered our mastery over other elements. Early humans used fire to shape tools, hardening wooden and stone implements. This innovation enabled our ancestors to extend their range, explore new territories, and adapt to different environments.

Fire as a Weapon:

Beyond its constructive uses, fire also became a potent weapon. Fire could be employed for hunting, warfare, and defence. Controlling fire gave early humans a critical advantage in their interactions with the natural world and with other species.

Fire as a Symbol and Ritual:

As civilizations developed, fire took on symbolic and ritualistic significance. It became associated with religious ceremonies, representing purification, transformation, and a connection to the divine. Flames illuminated temples, altars, and sacred spaces across cultures and ages.

Conclusion:

The harnessing of fire was a transformative event in human history, marking the beginning of our ascent to dominance on Earth. It provided us with warmth, nutrition, protection, and the means to shape the world around us. The mastery of fire was not only a physical achievement but also a symbol of our capacity to innovate and adapt.

Chapter 3
The Printing Press

Introduction:

The invention of the printing press by Johannes Gutenberg in the 15th century is widely regarded as one of the most transformative events in human history. In this chapter, we explore how the printing press revolutionized the dissemination of knowledge and ushered in an era of profound change.

Gutenberg's Ingenious Invention:

Johannes Gutenberg's invention of the printing press, around 1440 in Mainz, Germany, was a breakthrough that would forever alter the way information was shared. At its core, the printing press involved movable type—a system of individual letters and characters that could be rearranged and reused to print multiple copies of a text.

The Impact of the Gutenberg Bible:

The first major work produced on Gutenberg's press was the Bible, known as the Gutenberg Bible or the 42-line Bible. This monumental achievement made the Bible more accessible than ever before, shifting the balance of power in the church and paving the way for religious reformations.

The Democratization of Knowledge:

Prior to the printing press, books were painstakingly copied by hand, a laborious and time-consuming process that limited the availability of written knowledge to a privileged few. The printing press democratized knowledge, making books more affordable and accessible to a wider audience.

The Gutenberg Revolution:

The rapid dissemination of information led to a period known as the Gutenberg Revolution. It ignited a thirst for knowledge, fuelling the Renaissance and the Age of Enlightenment. Universities flourished, and ideas circulated freely.

The Spread of Scientific Inquiry:

The printing press played a pivotal role in the Scientific Revolution by allowing scientists to share their findings and collaborate across vast distances. It laid the foundation for modern science and empirical inquiry.

Political and Social Impact:

The printing press also had profound political and social consequences. It facilitated the spread of revolutionary ideas, including those that led to the American and French Revolutions. It challenged established authority and promoted the concept of individual rights and freedoms.

Cultural Diversity and Linguistic Identity:

Printing in various languages became possible, preserving and promoting linguistic diversity. It bolstered cultural identities and helped standardize languages.

The Legacy of the Printing Press:

The printing press left an indelible mark on human history, transforming education, culture, religion, and politics. It accelerated the pace of progress, making information widely available and fostering a spirit of inquiry that continues to shape our world.

Conclusion:

The printing press, with its revolutionary technology of movable type, shattered the barriers to knowledge and information dissemination. It empowered individuals, sparked intellectual revolutions, and transformed societies.

Chapter 4
Navigating with the Compass

Introduction:

The magnetic compass, a simple yet ingenious invention, has played a pivotal role in shaping the course of human exploration and navigation. This chapter explores how this unassuming device revolutionized our ability to traverse the globe and discover new horizons.

Early Navigational Challenges:

Before the compass, sailors and explorers faced daunting challenges when navigating the open seas. They relied on landmarks, stars, and dead reckoning, often resulting in perilous journeys and uncertainty about their positions.

The Birth of the Magnetic Compass:

The origins of the magnetic compass are shrouded in mystery, but it likely emerged in China during the Han Dynasty (2nd century BC). Early compasses were simple lodestones mounted on floating objects. These early devices could align with the Earth's magnetic field, providing a basic sense of direction.

The Mariner's Compass:

The magnetic compass evolved into the mariner's compass, a more refined and practical tool for navigation. By the 12th century, the compass had found its way to Europe, where it played a pivotal role in the Age of Exploration.

Era of Exploration:

The magnetic compass became indispensable for explorers like Christopher Columbus, Ferdinand Magellan, and Vasco da Gama. It allowed them to venture into uncharted waters, cross oceans, and circumnavigate the globe with greater confidence.

The Role of the North Star:

Navigators used the compass in conjunction with celestial navigation, often relying on the North Star (Polaris) as a reference point. This combination of tools enabled accurate and reliable positioning during long sea voyages.

Scientific Understanding of Magnetism:

Over time, scientists began to unravel the mysteries of magnetism, deepening our understanding of how the compass worked. This knowledge not only improved compass design but also contributed to the development of other magnetic instruments.

Impact Beyond Navigation:

The compass's influence extended beyond exploration. It played a crucial role in cartography, allowing for the creation of accurate maps. It also had applications in land navigation, contributing to the development of overland trade routes.

The Modern Compass:

Today, the magnetic compass is a ubiquitous tool used in navigation, from hiking and orienteering to aviation and maritime travel. Its principles are still relevant in a world filled with advanced technology.

Conclusion:

The magnetic compass, a seemingly simple device, transformed the art of navigation and exploration. It empowered adventurers to cross vast oceans, discover new lands, and connect distant civilizations.

Chapter 5
The Steam Engine

Introduction:

The steam engine, a marvel of engineering ingenuity, stands as the driving force behind the Industrial Revolution. In this chapter, we explore how this remarkable invention transformed industry, transportation, and society as a whole.

The Birth of Steam Power:

The concept of harnessing steam as a source of power dates back to ancient times, but it was during the 18th century that engineers like Thomas Savery and Thomas Newcomen began developing rudimentary steam engines. These early machines were used primarily to pump water from mines.

James Watt's Innovation:

The real turning point came with the improvements made by Scottish engineer James Watt in the late 18th century. Watt's innovative design, featuring a separate condenser, significantly increased the efficiency of steam engines, making them practical for a wide range of applications.

Revolutionizing Industry:

The steam engine revolutionized industry by providing a reliable and powerful source of mechanical energy. It replaced waterwheels and horse-drawn machinery in factories, leading to a surge in production capacity. Factories could be located anywhere, not just near water sources, giving rise to urban industrial centres.

Transportation Revolution:

Steam power also had a profound impact on transportation. The development of steam locomotives, such as George Stephenson's

"Rocket," enabled the rapid expansion of railways. Steamships transformed long-distance travel and trade, reducing travel times across oceans.

The Textile Industry and Beyond:

The textile industry was one of the first sectors to benefit from steam power, with machines like the cotton gin and spinning jenny increasing efficiency. Steam engines also drove the mining industry, allowing for the extraction of coal and minerals on a massive scale.

Urbanization and Social Change:

The rise of steam-powered industry led to significant urbanization as people moved from rural areas to cities in search of work. This shift had profound social and cultural consequences, shaping the modern cityscape and workforce.

Environmental and Economic Impact:

The Industrial Revolution, fuelled by steam power, had both positive and negative consequences. While it spurred economic growth and technological advancement, it also brought about environmental challenges and social inequalities.

Legacy and Technological Progress:

The steam engine's legacy extends beyond the 18th and 19th centuries. It laid the groundwork for subsequent innovations in energy production and transportation, including the development of the internal combustion engine and electricity generation.

Conclusion:

The steam engine, once a humble device used to pump water from mines, evolved into a transformative force that powered the Industrial Revolution. It reshaped industry, transportation, and society, leaving an indelible mark on human progress.

Chapter 6
The Telegraph

Introduction:

The telegraph, a ground-breaking communication technology, revolutionized the way people exchanged information and bridged vast distances. In this chapter, we delve into the story of the telegraph and its role in connecting continents through the use of Morse code.

The Need for Faster Communication:

Before the telegraph, long-distance communication relied on methods like letters, messengers, and semaphore systems. These methods were slow and often unreliable, particularly for communicating urgent information.

Samuel Morse and the Telegraph:

The invention of the telegraph is closely associated with Samuel Morse, an American inventor and artist. In the early 1830s, Morse conceived the idea of using electrical signals to transmit messages over long distances.

Morse Code: The Language of the Telegraph:

To make long-distance communication efficient, Morse developed a code that represented letters and numbers using combinations of dots and dashes. This code, known as Morse code, became the universal language of telegraphy and is still used in various forms of communication today.

The First Telegraph Line:

In 1844, Morse successfully demonstrated his telegraph system by sending the first telegraph message, "What hath God wrought," over a line from Washington, D.C., to Baltimore. This achievement marked the birth of a new era in communication.

Rapid Expansion:

The telegraph spread rapidly across the United States and Europe. Telegraph lines crisscrossed continents, connecting major cities and towns. Messages that once took weeks or months to deliver could now be transmitted almost instantly.

Impact on Commerce and Society:

The telegraph had a profound impact on commerce, allowing businesses to make quicker decisions based on up-to-date information. It transformed journalism, enabling newspapers to report news from distant locations promptly.

Global Communication: Transatlantic Telegraph:

One of the most significant achievements in telegraphy was the laying of the transatlantic telegraph cable in 1858. This underwater cable connected North America and Europe, allowing for near-instant communication between the continents.

Challenges and Resilience:

The telegraph faced challenges, including technical difficulties and competition, but it continued to evolve. Improvements in cable design, transmission speed, and telegraphy equipment enhanced its capabilities.

Legacy of the Telegraph:

The telegraph laid the groundwork for future communication technologies, including the telephone and the internet. Its impact on global communication and the exchange of information cannot be overstated.

Conclusion:

The telegraph, with its Morse code language and rapid message transmission, transformed long-distance communication and brought the world closer together. It was a testament to human ingenuity and persistence, paving the way for the interconnected world we know today.

Chapter 7
The Telephone

Introduction:

The telephone, one of the most transformative inventions of the 19th century, forever altered the way people communicated. In this chapter, we delve into the fascinating story of the telephone and how it bridged voices across vast distances.

The Need for Instant Communication:

Before the telephone, communication over long distances was limited to written messages, telegrams, or face-to-face meetings. The desire for instant voice communication was a long-standing aspiration.

Alexander Graham Bell's Invention:

In 1876, Alexander Graham Bell, a Scottish-born inventor residing in the United States, was awarded the patent for the telephone. Bell's invention allowed for the transmission of spoken words over electrical wires.

The First Telephone Call:

The historic first telephone call took place on March 10, 1876, when Bell spoke to his assistant, Thomas Watson, in another room, saying, "Mr. Watson, come here, I want to see you." It was a moment that marked the birth of the telephone era.

Early Challenges and Improvements:

The early telephone technology had its limitations, including sound quality and range. However, inventors and engineers tirelessly worked to refine and expand the capabilities of the telephone network.

The Telephone Network Expands:

The telephone network rapidly expanded, connecting homes, businesses, and communities. Telephone exchanges and switchboards facilitated the routing of calls, enabling people to connect with others across towns, states, and eventually countries.

Impact on Communication:

The telephone transformed the way people communicated. It allowed for immediate and direct conversations, transcending the constraints of time and distance. Families could stay in touch, businesses could conduct transactions, and emergency services could respond more effectively.

Cultural and Social Impact:

The telephone influenced culture and society. It brought people closer together, fostering connections and relationships. It also played a role in shaping language and communication etiquette.

Technological Advancements:

Advancements in telephone technology led to innovations like the rotary dial, which replaced manual operator connections, and later, push-button dialling. The introduction of long-distance lines and transatlantic cables expanded the reach of the telephone network.

The Mobile Phone Era:

In the late 20th century, the telephone underwent another revolution with the advent of mobile phones. These portable devices allowed for communication on the go, further enhancing connectivity.

Legacy of the Telephone:

The telephone laid the foundation for modern telecommunications. It paved the way for subsequent inventions, such as the internet and smartphones, which have taken communication to new heights.

Conclusion:

The telephone, a marvel of 19th-century innovation, broke down barriers of time and distance, allowing people to connect in ways previously unimaginable. It transformed the way we communicate, conduct business, and interact with one another, leaving an indelible mark on human history.

Chapter 8
Electricity

Introduction:

Electricity, one of the most fundamental forces of nature, has been harnessed to revolutionize nearly every aspect of modern life. In this chapter, we delve into the captivating story of electricity and how it has illuminated and powered our world.

The Nature of Electricity:

Electricity has been a part of our natural world since ancient times. The Greeks and Romans were aware of static electricity, generated by rubbing materials like amber and fur. However, understanding and harnessing the full potential of electricity would require centuries of scientific inquiry.

The Age of Enlightenment:

The 18th century witnessed significant discoveries in the realm of electricity. Benjamin Franklin's kite experiment helped establish the link between lightning and electricity. Meanwhile, Alessandro Volta invented the first chemical battery, known as the Voltaic pile, in 1800.

Michael Faraday and Electromagnetism:

The 19th century saw pivotal breakthroughs in understanding electricity. British scientist Michael Faraday's work on electromagnetism laid the foundation for modern electrical technology. He discovered electromagnetic induction, which demonstrated the generation of electricity through the motion of a conductor within a magnetic field.

Thomas Edison and the Electric Light Bulb:

One of the most iconic applications of electricity was the invention of the incandescent light bulb. Thomas Edison's development of a practical, long-lasting bulb in the late 19th century marked a turning point in history, illuminating cities and homes.

The Power of Alternating Current (AC):

The "War of the Currents" between Edison, who championed direct current (DC), and Nikola Tesla and George Westinghouse, who promoted alternating current (AC), shaped the electrical landscape. AC ultimately prevailed due to its ability to transmit electricity over longer distances, leading to the creation of extensive power grids.

Electrifying Industries:

Electricity revolutionized industries by providing a versatile source of power. Factories, mines, and transportation systems adopted electric motors, increasing efficiency and productivity.

Household Electrification:

The widespread adoption of electricity in homes transformed daily life. Electric appliances, from refrigerators to washing machines, made household chores more manageable and improved living standards.

Technological Advancements:

The development of electrical components and devices, including transformers, capacitors, and transistors, expanded the possibilities of electrical engineering. These innovations paved the way for the digital age and modern electronics.

Renewable Energy and Sustainability:

In recent years, there has been a growing emphasis on harnessing electricity from renewable sources, such as solar and wind power, to address environmental concerns and reduce dependence on fossil fuels.

Conclusion:

Electricity, once a mysterious force of nature, has become the lifeblood of the modern world. It has not only illuminated our nights but also powered our industries, connected our societies, and driven technological progress.

Chapter 9
The Light Bulb

Introduction:

The invention of the incandescent light bulb by Thomas Edison in the late 19th century marked a historic moment in human history. In this chapter, we explore the fascinating story of the light bulb and how it illuminated the world.

The Need for Artificial Light:

Throughout history, people had relied on various sources of artificial light, including candles, oil lamps, and gas lamps. However, these sources had limitations in terms of brightness, safety, and convenience.

Edison's Determination:

Thomas Edison, an American inventor known for his tenacity and innovation, set out to create a practical and long-lasting electric light bulb. His goal was to provide a reliable source of illumination for homes and businesses.

The Quest for the Right Filament:

Edison and his team conducted countless experiments to find the perfect filament material that would glow brightly when heated by an electric current. After testing thousands of materials, they settled on a carbonized bamboo filament.

The First Practical Incandescent Bulb:

In 1879, Edison unveiled the first practical incandescent light bulb. It consisted of a thin filament encased in a vacuum-sealed glass bulb. When an electric current passed through the filament, it glowed brilliantly, producing a steady and reliable source of light.

The Impact on Society:

The introduction of the incandescent light bulb revolutionized society. It extended the hours of productivity and leisure, transforming how people lived and worked. Cities became vibrant with night-time activity, and factories could operate around the clock.

Economic and Cultural Transformation:

Electric lighting had a profound economic impact, reducing the cost of illumination and increasing energy efficiency. It also influenced cultural norms, changing the way people experienced their environments and entertainment.

Urbanization and City Planning:

The widespread adoption of electric lighting influenced urban planning and architecture. Streets were illuminated, making them safer and more accessible at night. Skyscrapers and neon signs became iconic symbols of modern cities.

Technological Advancements:

The development of the light bulb spurred advancements in electrical engineering, including improvements in electrical grids, power distribution, and lighting fixtures. It also laid the foundation for future innovations in lighting technology.

Sustainability and LED Lighting:

In recent years, there has been a shift toward more energy-efficient lighting technologies, such as light-emitting diodes (LEDs). These innovations offer greater energy savings and reduced environmental impact.

Conclusion:

The incandescent light bulb, born out of Edison's determination and relentless experimentation, brightened the world and transformed the way we live and work. It illuminated our nights, extended our productivity, and shaped the modern urban landscape.

Chapter 10
The Internal Combustion Engine

Introduction:

The internal combustion engine stands as one of the most transformative inventions in the history of transportation and industry. In this chapter, we delve into the captivating story of the internal combustion engine and its role in driving innovation across various domains.

The Early Days of Engine Development:

The concept of harnessing the power of explosions to generate mechanical work dates back centuries. Early inventors experimented with various engine designs, but it was in the late 19th century that the internal combustion engine began to take shape.

The Four-Stroke Engine:

Nikolaus Otto, a German engineer, is credited with developing the first practical internal combustion engine in 1876. His four-stroke engine, also known as the Otto engine, introduced a systematic approach to combustion engines, featuring intake, compression, power, and exhaust strokes.

Henry Ford and the Automobile Revolution:

The widespread adoption of the internal combustion engine was closely tied to the rise of the automobile industry. Henry Ford's introduction of the assembly line in the early 20th century made cars more affordable and accessible, revolutionizing personal transportation.

Impact on Transportation:

The internal combustion engine rapidly replaced steam engines and horse-drawn carriages in the transportation sector. It enabled faster, more reliable, and longer-distance travel, transforming the movement of goods and people.

Industrial Applications:

The internal combustion engine found applications beyond automobiles. It powered tractors, trucks, ships, and airplanes, enhancing the efficiency and productivity of industries worldwide. It also facilitated the development of heavy machinery and construction equipment.

Military Advancements:

During World War I and World War II, internal combustion engines played a pivotal role in military vehicles and aircraft. They provided the mobility and power needed for modern warfare.

Innovations in Fuel:

The internal combustion engine prompted innovations in fuel development, from gasoline to diesel and alternative fuels. These advances addressed efficiency, emissions, and sustainability concerns.

Environmental and Efficiency Challenges:

While internal combustion engines offered tremendous benefits, they also raised environmental challenges, including air pollution and reliance on fossil fuels. Research into fuel efficiency and emissions control became critical.

Hybrid and Electric Technologies:

In response to environmental concerns, hybrid and electric propulsion systems have emerged as alternatives to traditional internal combustion engines. These technologies aim to reduce emissions and dependence on fossil fuels.

Global Impact:

The internal combustion engine has had a profound global impact, shaping economies, urban planning, and consumer lifestyles. It has fostered mobility, trade, and globalization.

Continued Innovation:

Advancements in internal combustion engine technology continue with a focus on efficiency, performance, and cleaner emissions. Engineers are exploring turbocharging, direct injection, and advanced materials to enhance engine design.

Conclusion:

The internal combustion engine, with its capacity to convert fuel into mechanical energy, has driven innovation across transportation, industry, and warfare. It has reshaped societies, economies, and the way people live and work.

Chapter 11
The Airplane

Introduction:

The invention of the airplane represents one of humanity's most extraordinary achievements, marking the conquest of the skies and transforming the way we travel and explore. In this chapter, we delve into the captivating story of the airplane and its role as the wings of progress.

Dreams of Flight:

For centuries, humans dreamt of soaring through the air like birds. Early inventors designed ornithopters and gliders, attempting to unlock the secrets of controlled flight. However, it was the Wright brothers who would achieve this dream.

The Wright Brothers' Breakthrough:

In December 1903, Orville and Wilbur Wright achieved powered, controlled, and sustained flight with their Wright Flyer in Kitty Hawk, North Carolina. This momentous achievement marked the birth of the modern airplane.

The Pioneering Years:

The early years of aviation were marked by daring aviators and rapid technological advancements. Innovations in aircraft design, engine development, and navigation systems paved the way for the expansion of aviation.

World War I and Military Aviation:

World War I saw the first widespread use of airplanes in warfare. They were used for reconnaissance, aerial combat, and bombardment. This period accelerated aircraft technology and tactics.

The Golden Age of Aviation:

The post-World War I era witnessed the Golden Age of Aviation. Aviators like Charles Lindbergh and Amelia Earhart captured the world's imagination with record-breaking flights and transatlantic crossings.

Commercial Aviation Takes Flight:

The 1920s and 1930s marked the dawn of commercial aviation. Airlines began offering passenger services, connecting cities and countries. Innovations like pressurized cabins and larger aircraft made long-distance travel more comfortable.

World War II and Technological Advancements:

World War II brought further innovations in aviation, including faster and more powerful aircraft like the jet engine-powered fighters. The war also saw the development of strategic bombing and long-range reconnaissance.

The Jet Age:

The post-World War II era ushered in the Jet Age. Jet airliners like the Boeing 707 and Douglas DC-8 made air travel faster and more accessible to the masses. This period also saw the advent of supersonic flight.

Space Exploration and the Space Race:

The same technology that propelled airplanes into the stratosphere paved the way for space exploration. The Space Race between the United States and the Soviet Union led to significant advancements in rocketry and spaceflight.

Modern Aviation:

Today, aviation is an integral part of global transportation and commerce. Commercial airliners crisscross the world, connecting people and goods. Advances in materials, avionics, and safety systems continue to enhance the industry.

Environmental and Sustainability Challenges:

While aviation has made the world more accessible, it also faces environmental challenges related to emissions and sustainability. Research into alternative fuels and cleaner aircraft technologies is ongoing.

The Future of Flight:

The future of aviation holds promises of supersonic travel, electric propulsion, and even space tourism. Engineers and innovators continue to push the boundaries of what is possible in flight.

Conclusion:

The airplane, as the wings of progress, has revolutionized how we explore, travel, and connect with the world. From the pioneering efforts of the Wright brothers to the modern age of aviation, it has been a symbol of human achievement and a catalyst for technological advancement.

Chapter 12
Penicillin

Introduction:

The discovery of penicillin, the first antibiotic, was a ground-breaking moment in the history of medicine. In this chapter, we delve into the fascinating story of penicillin and its transformative impact on healthcare and human well-being.

The Pre-Antibiotic Era:

Before the discovery of antibiotics, bacterial infections were a significant threat to human health. Common infections, such as pneumonia, strep throat, and skin infections, often turned deadly, and surgeries carried a high risk of infection.

Alexander Fleming's Discovery:

The story of penicillin begins with the accidental discovery by Scottish biologist and pharmacologist Alexander Fleming in 1928. While working with Staphylococcus bacteria, Fleming noticed that a mould called Penicillium notatum produced a substance that killed the bacteria.

The Development of Penicillin:

Fleming's discovery laid the foundation for the development of penicillin as a therapeutic agent. However, it wasn't until the 1940s, during World War II, that scientists, including Howard Florey and Ernst Boris Chain, successfully purified and mass-produced penicillin for medical use.

The Antibiotic Revolution:

The introduction of penicillin and subsequent antibiotics marked a revolution in medicine. For the first time, bacterial infections could be effectively treated with a simple and safe medication.

Impact on Medicine and Surgery:

Penicillin transformed the practice of medicine and surgery. Surgeries that were once too risky due to infection risk became safer. Doctors could treat life-threatening infections, saving countless lives.

The Era of Antibiotics:

The success of penicillin spurred the discovery and development of numerous other antibiotics, expanding the medical toolkit for combating bacterial diseases.

Challenges of Antibiotic Resistance:

Over time, the misuse and overuse of antibiotics led to the emergence of antibiotic-resistant bacteria. Antibiotic resistance remains a significant global health challenge, emphasizing the importance of responsible antibiotic use.

Broad Applications:

Antibiotics found applications not only in human medicine but also in veterinary medicine and agriculture. They played a vital role in ensuring food safety and animal health.

Global Impact:

Antibiotics have had a profound global impact, reducing mortality from infectious diseases and improving overall public health. They have been instrumental in the control of epidemics and pandemics.

Sustainable Antibiotic Use:

The responsible use of antibiotics and ongoing research into new antibiotic classes are essential for preserving their effectiveness and addressing antibiotic resistance.

Conclusion:

Penicillin, the first antibiotic, and subsequent antibiotics have revolutionized medicine, saving countless lives and reshaping our approach to bacterial infections. They are a testament to human ingenuity and the power of scientific discovery

Chapter 13
The Internet

Introduction:

The internet, often referred to as the global village, has emerged as a transformative force, reshaping how we communicate, access information, and connect with the world. In this chapter, we delve into the captivating story of the internet and its role in connecting the global village.

The World as a Global Village:

The term "global village" was popularized by media theorist Marshall McLuhan in the 1960s, envisioning a world where technology connects people across vast distances. The internet has turned this vision into reality.

The Internet's Origins:

The internet's roots can be traced back to the Cold War era when the U.S. Department of Defence developed ARPANET, a research network designed to withstand nuclear attacks. ARPANET laid the foundation for the internet as we know it.

The World Wide Web:

Tim Berners-Lee's invention of the World Wide Web in 1989 was a pivotal moment. It introduced a user-friendly system for organizing and accessing information, making the internet accessible to people worldwide.

Expanding Global Connectivity:

The 1990s saw the internet expand beyond academia and the military to reach the public. The rise of web browsers, email services, and internet service providers made connectivity more accessible.

The Dot-Com Boom and E-Commerce:

The late 1990s witnessed the dot-com boom, with companies like Amazon and eBay emerging as online retail giants. E-commerce revolutionized the way people shop and conduct business.

Social Media and Online Communities:

The 2000s saw the rise of social media platforms like Facebook, Twitter, and YouTube. These platforms transformed how people connect, share information, and engage with global communities.

Information Access and Knowledge Sharing:

The internet democratized access to information and facilitated knowledge sharing. Platforms like Wikipedia and open-access journals have empowered individuals and researchers.

The Mobile Internet:

The introduction of smartphones and mobile devices brought the internet into the palms of people worldwide. Mobile internet connectivity and the development of mobile apps transformed how we access information and interact online.

Global Challenges and Opportunities:

The internet has brought both opportunities and challenges. Issues like cybersecurity, privacy concerns, and digital divides require ongoing attention and innovation.

Digital Economies and Globalization:

The internet has facilitated global trade, remote work, and digital economies. It has reshaped industries and allowed businesses to reach international markets.

The Internet as a Tool for Activism:

The internet has also played a significant role in social and political activism. Online movements and campaigns have raised awareness, mobilized communities, and driven social change.

The Future of the Internet:

The future holds promise of increased connectivity through 5G and satellite networks, as well as innovations like the Internet of Things (IoT) and artificial intelligence (AI) that will further transform how we interact with technology.

Conclusion:

The internet, as a tool for connecting the global village, has redefined the way we live, work, and communicate. It has created opportunities for collaboration, cultural exchange, and the dissemination of ideas on a scale previously unimaginable.

Chapter 14
The Computer

Introduction:

The invention of the computer ushered in the age of information technology, transforming the way we process data, solve complex problems, and communicate. In this chapter, we delve into the captivating story of the computer and its role in shaping the modern world.

The Emergence of Computing Machines:

The roots of computing can be traced back to ancient civilizations like the Greeks and the Chinese who developed rudimentary calculating devices. However, the modern computer as we know it began to take shape in the 20th century.

The Birth of the Electronic Computer:

The electronic computer era started in the 1940s with the development of machines like the ENIAC and UNIVAC. These early computers were enormous and used vacuum tubes for processing.

The Turing Machine and Theory of Computation:

Theoretical work by mathematicians like Alan Turing and John von Neumann laid the foundation for computer science and the concept of a universal machine capable of performing any computation.

The Transistor Revolution:

The invention of the transistor in the 1950s marked a pivotal moment. Transistors replaced vacuum tubes, making computers smaller, more reliable, and more energy-efficient. This led to the development of mainframes and minicomputers.

The Personal Computer Revolution:

The 1970s and 1980s witnessed the emergence of personal computers, with pioneers like Steve Jobs and Bill Gates leading the way. The IBM PC and the Apple Macintosh brought computing into homes and offices.

The Internet and Networking:

The integration of computers and networks in the 1980s paved the way for the internet's expansion. The World Wide Web, email, and online services transformed communication and information access.

Software and Operating Systems:

The development of software and operating systems, including Unix, Windows, and Linux, made computers more user-friendly and versatile. Software applications diversified, catering to various needs and industries.

Artificial Intelligence and Machine Learning:

Advancements in artificial intelligence (AI) and machine learning have pushed the boundaries of what computers can do. AI-driven technologies like speech recognition, image analysis, and autonomous systems are becoming increasingly integrated into our lives.

Big Data and Data Analytics:

The digital age has generated vast amounts of data. Computers equipped with powerful processors and data analytics tools enable organizations to extract insights and make data-driven decisions.

Cybersecurity and Privacy:

As the role of computers expanded, so did concerns about cybersecurity and privacy. Cyberattacks, data breaches, and the need for secure digital identities have become critical issues.

The Computer in Science and Research:

Computers have become indispensable tools in scientific research, from simulations in physics to DNA sequencing in biology. They enable complex calculations and simulations that were once inconceivable.

The Internet of Things (IoT) and Connectivity:

The IoT has connected everyday objects to the internet, creating smart homes, cities, and industries. This interconnectivity is revolutionizing sectors like healthcare, transportation, and agriculture.

Quantum Computing and Future Horizons:

Quantum computing, still in its early stages, promises to revolutionize computing once again with the potential to solve complex problems at speeds unattainable by classical computers.

Conclusion:

The computer, as the cornerstone of information technology, has redefined the way we live, work, and interact with the world. Its evolution continues to shape the modern age, fostering innovation, scientific discovery, and unprecedented connectivity.

Chapter 15
The Smartphone

Introduction:

The smartphone has emerged as a transformative invention, revolutionizing how we communicate, access information, entertain ourselves, and conduct business in a compact, pocket-sized device. In this chapter, we delve into the captivating story of the smartphone and its role as a pocket-sized revolution.

Early Mobile Phones:

The concept of mobile communication predates the smartphone by several decades. Early mobile phones were large and bulky, primarily used for voice calls, and were a luxury available to only a few.

The Birth of the Smartphone:

The term "smartphone" began to gain prominence in the early 21st century with the launch of devices like the BlackBerry and the Palm Treo. These phones offered features beyond voice calls, such as email and basic internet browsing.

The Apple iPhone:

In 2007, Apple revolutionized the mobile industry with the introduction of the iPhone. Combining a touchscreen interface, internet connectivity, multimedia capabilities, and third-party apps, the iPhone set a new standard for smartphones.

Android and the Rise of Competing Platforms:

Google's Android operating system provided a flexible and open-source alternative to iOS, leading to the proliferation of Android smartphones from various manufacturers. This competition spurred innovation and expanded consumer choice.

Mobile Apps and Ecosystems:

The App Store and Google Play Store revolutionized how software is distributed and consumed. The availability of a vast array of mobile apps for productivity, entertainment, and utilities has become a hallmark of smartphones.

Connectivity and the Mobile Internet:

Smartphones brought the internet to users' fingertips, enabling email, web browsing, and social media access on the go. High-speed mobile networks like 4G and 5G further enhanced connectivity.

Multimedia and Entertainment:

The integration of high-quality cameras, music players, and video streaming capabilities turned smartphones into versatile multimedia devices. They revolutionized photography, video recording, and content consumption.

Productivity and Business Tools:

Smartphones have become indispensable tools for productivity and business. Features like email, calendars, document editing, and video conferencing facilitate remote work and collaboration.

Social Media and Digital Communication:

Social media platforms and instant messaging apps have reshaped how people connect and communicate. Smartphones have become the primary means of staying in touch with friends and family.

Mobile Payments and Digital Wallets:

The adoption of mobile payments and digital wallets has transformed the way we conduct financial transactions, making purchases, and transferring money using smartphones.

Health and Fitness:

Smartphones offer a range of health and fitness tracking apps, enabling users to monitor their physical activity, nutrition, and overall well-being. They have also become tools for telemedicine and remote healthcare.

Challenges of Smartphone Use:

The widespread use of smartphones has raised concerns about screen addiction, privacy, cybersecurity, and digital well-being. Researchers and experts continue to explore ways to address these challenges.

The Future of Smartphones:

The evolution of smartphones continues with advancements like foldable displays, augmented reality (AR), and enhanced artificial intelligence (AI) integration. These technologies promise to further transform how we interact with our devices.

Conclusion:

The smartphone, as a pocket-sized revolution, has fundamentally changed the way we live, work, and communicate. It has consolidated a wide range of functions into a single device, shaping our daily routines and empowering us with unprecedented connectivity and convenience.

Chapter 16
The Radio

Introduction:

The invention of the radio marked a profound shift in how people received information and entertainment, bringing voices and music from afar directly into their homes. In this chapter, we delve into the captivating story of the radio and its role as a medium for voices from afar.

The Pre-Radio Era:

Before the radio, communication was limited to face-to-face interactions, written correspondence, and printed publications. The idea of transmitting sound through electromagnetic waves was a ground-breaking concept.

The Pioneers of Radio:

The development of the radio was the collective effort of inventors, scientists, and engineers. Key figures like Guglielmo Marconi, Nikola Tesla, and Heinrich Hertz made significant contributions to the understanding of wireless communication.

Marconi's Transatlantic Success:

In 1901, Guglielmo Marconi achieved a historic milestone by sending the first wireless transatlantic message from England to Newfoundland. This achievement laid the foundation for long-distance radio communication.

The Birth of Commercial Radio:

The 1920s saw the emergence of commercial radio broadcasting. Radio stations began to transmit music, news, and entertainment to a growing audience. The BBC in the United Kingdom and KDKA in the United States were among the early pioneers.

Radio's Role in World War I and II:

During both World Wars, radio played a crucial role in disseminating news and propaganda. It connected soldiers with their families and served as a tool for diplomacy and intelligence.

Golden Age of Radio:

The 1930s and 1940s are often referred to as the Golden Age of Radio. Radio shows, such as "The War of the Worlds," "The Shadow," and "The Lone Ranger," captivated audiences and became cultural phenomena.

Breaking Barriers:

Radio became a platform for breaking societal and cultural barriers. It allowed voices and music from diverse backgrounds to reach a wide audience, fostering a sense of shared experience and unity.

News and Information Dissemination:

Radio played a pivotal role in disseminating news during major events like the Hindenburg disaster, the coronation of Queen Elizabeth II, and the attack on Pearl Harbour. It provided real-time updates to the public.

Music and Entertainment:

Radio introduced people to new genres of music, from jazz to rock 'n' roll. It also offered a wide range of entertainment, including radio dramas, comedy shows, and variety programs.

Challenges and Adaptations:

The advent of television in the mid-20th century posed a challenge to radio, but the medium adapted by focusing on music, talk shows, and niche programming.

Community and Local Radio:

Local and community radio stations have continued to play a vital role in broadcasting news, culture, and events specific to their regions.

Technological Advancements:

The transition from analogue to digital broadcasting and the development of internet radio have expanded the reach and capabilities of radio.

Radio in the Digital Age:

Today, radio remains a relevant and accessible medium, with online streaming, podcasts, and satellite radio offering diverse content to global audiences.

Conclusion:

The radio, as a medium for voices from afar, has connected people across vast distances, bringing them news, music, and entertainment. It has played a significant role in shaping culture, society, and the way we receive information and inspiration.

Chapter 17
The Television

Introduction:

The television, often referred to as the "window to the world," has had a profound impact on society, culture, and the way we consume information and entertainment. In this chapter, we delve into the captivating story of the television and its role as a window to the world.

Pre-Television Era:

Before the television, people relied on newspapers, radio broadcasts, and cinema for news and entertainment. The idea of transmitting moving images through the air was a revolutionary concept.

Early Experiments and Inventors:

Inventors and scientists like John Logie Baird, Philo Farnsworth, and Vladimir Zworykin conducted pioneering experiments in television technology during the early 20th century.

The Emergence of Television Broadcasting:

In the 1920s and 1930s, television broadcasting began in experimental form. Public demonstrations and broadcasts of events like the 1936 Berlin Olympics introduced the world to television.

The First Televised Programs:

The late 1930s saw the launch of regular television programming. Early shows included newsreels, variety programs, and experimental dramas.

World War II and Television:

During World War II, television technology advanced significantly. Television broadcasts of the war effort and news coverage became important sources of information and propaganda.

Post-War Prosperity and the Rise of the Television Set:

The post-war era brought economic prosperity, leading to the mass production and adoption of television sets in households. The 1950s became known as the "golden age of television."

Iconic Shows and Cultural Impact:

Television shows like "I Love Lucy," "The Twilight Zone," and "The Ed Sullivan Show" became cultural phenomena. They brought entertainment and shared experiences into living rooms across the nation.

The Role of Television in Civil Rights and Social Change:

Television played a crucial role in broadcasting civil rights struggles, such as the coverage of the Montgomery Bus Boycott and Martin Luther King Jr.'s speeches. It helped raise awareness and mobilize support for social change.

News and Information:

Television became a primary source of news with programs like "The Huntley-Brinkley Report" and Walter Cronkite's coverage of historic events such as the moon landing and the Vietnam War.

Television as an Educational Tool:

Educational programs like "Sesame Street" and "The Discovery Channel" demonstrated the educational potential of television, teaching children and adults alike.

Cable and Satellite Television:

The introduction of cable and satellite television in the late 20th century expanded the number of available channels, offering specialized content and global reach.

Technological Advancements:

The transition from analogue to digital broadcasting and the development of high-definition (HD) and smart TVs have improved the quality and interactivity of television.

Television in the Digital Age:

The digital age has brought streaming services, on-demand content, and the ability to watch television on multiple devices. This has transformed how we consume and interact with television.

Challenges and Opportunities:

Television faces challenges from online streaming and cord-cutting trends. However, it continues to evolve with innovations like 4K resolution and interactive content.

Conclusion:

The television, as a window to the world, has brought information, entertainment, and cultural experiences into the homes of billions. It has shaped public perception, influenced social change, and provided a platform for creativity and storytelling.

Chapter 18
The Transistor

Introduction:

The transistor, a tiny yet powerful electronic component, stands as one of the most influential inventions of the 20th century. It has revolutionized the field of electronics, enabling the miniaturization of devices and shaping the modern world. In this chapter, we delve into the captivating story of the transistor and its role as a miniaturization marvel.

The Vacuum Tube Era:

Before the transistor, electronic devices relied on vacuum tubes, which were large, fragile, and consumed significant power. These limitations hindered the development of portable and efficient electronics.

The Birth of the Transistor:

In 1947, engineers John Bardeen, Walter Brattain, and William Shockley at Bell Laboratories developed the first transistor, a semiconductor device capable of amplifying and switching electrical signals. This invention marked a paradigm shift in electronics.

How the Transistor Works:

Transistors are made from semiconductor materials, typically silicon or germanium. They can control the flow of electrical current by varying the voltage applied to them. This ability to amplify and switch signals is fundamental to modern electronics.

The Impact of the Transistor:

The transistor's impact on technology and society has been profound. It has led to the development of smaller, more efficient, and more reliable electronic devices.

The Rise of Miniaturization:

Transistors enabled the miniaturization of electronic components and devices. The size and power requirements of electronics dramatically decreased, making portable devices like radios, calculators, and eventually, smartphones, possible.

Computers and Digital Revolution:

Transistors played a pivotal role in the development of computers. The transition from vacuum tubes to transistors marked the beginning of the digital revolution, making computing more accessible and affordable.

Telecommunications and Modern Connectivity:

Transistors are fundamental to telecommunications equipment, including mobile phones and network infrastructure. They enable fast and efficient data processing and transmission.

Consumer Electronics:

The transistor's miniaturization impact extended to consumer electronics like televisions, radios, and audio players, making them more portable and energy-efficient.

Space Exploration and Aerospace:

Transistors' reliability and compactness made them crucial components in space exploration and aerospace technology, where size and weight constraints are critical.

Medicine and Healthcare:

Medical devices, from imaging equipment to implantable devices, rely on transistors for precision and efficiency. They have transformed diagnosis and treatment in healthcare.

Renewable Energy and Green Technology:

Transistors are integral to renewable energy systems like solar panels and wind turbines. They contribute to energy efficiency and the reduction of carbon emissions.

Challenges and Advances:

The ongoing challenge in transistor technology is to continue shrinking their size while maintaining performance. Advances like Moore's Law have guided the industry's progress.

Beyond Silicon:

Researchers are exploring alternative materials and approaches, such as quantum dots and carbon nanotubes, to push the boundaries of transistor technology.

Conclusion:

The transistor, as a miniaturization marvel, has reshaped our world, enabling the development of portable, powerful, and energy-efficient electronic devices. Its influence extends across industries, from communication to healthcare and beyond.

Chapter 19
The Nuclear Reactor

Introduction:

The nuclear reactor represents a revolutionary invention in the realm of energy production, offering the promise of abundant, clean, and efficient power. In this chapter, we delve into the captivating story of the nuclear reactor and its role in powering the future.

Early Discoveries in Nuclear Physics:

The groundwork for nuclear reactors was laid in the early 20th century with the discovery of nuclear phenomena such as radioactivity and the understanding of the structure of the atom.

The Manhattan Project and the Birth of Nuclear Energy:

During World War II, the Manhattan Project brought together leading scientists to develop nuclear weapons. This research also laid the foundation for harnessing nuclear energy for peaceful purposes.

The First Controlled Nuclear Chain Reaction:

In December 1942, physicist Enrico Fermi achieved the first controlled nuclear chain reaction at the University of Chicago. This breakthrough demonstrated the feasibility of controlled nuclear reactions for energy production.

The Dawn of Nuclear Power:

The 1950s marked the beginning of nuclear power generation with the construction of the Shipping port Atomic Power Station in Pennsylvania, the first full-scale nuclear power plant.

How Nuclear Reactors Work:

Nuclear reactors utilize the process of nuclear fission, where the nucleus of an atom is split into smaller fragments, releasing a tremendous amount of energy. This energy is harnessed to produce heat, which, in turn, generates electricity.

Nuclear Energy's Advantages:

Nuclear energy offers several advantages, including the absence of greenhouse gas emissions during power generation, a stable and reliable energy source, and the potential for vast energy output.

Challenges and Concerns:

Despite its advantages, nuclear energy also presents challenges and concerns, such as nuclear accidents, radioactive waste disposal, and the risk of nuclear weapons proliferation.

Types of Nuclear Reactors:

There are various types of nuclear reactors, including pressurized water reactors (PWRs), boiling water reactors (BWRs), and fast breeder reactors (FBRs), each with its unique design and characteristics.

Nuclear Energy Around the World:

Nuclear power is used in many countries as a significant source of electricity. France, for example, relies heavily on nuclear energy for its power needs.

Nuclear Fusion: The Holy Grail of Energy: Nuclear fusion, the process of combining light atomic nuclei to release energy, has the potential to provide virtually limitless and cleaner energy. Research into controlled nuclear fusion continues worldwide.

Safety Measures and Regulation:

Stringent safety measures and regulations govern the operation of nuclear reactors to prevent accidents and protect both the environment and public health.

The Future of Nuclear Energy:

The future of nuclear energy includes advancements in reactor technology, such as small modular reactors (SMRs), which offer enhanced safety and flexibility, as well as research into advanced fuel cycles and waste management.

Global Energy Challenges:

Nuclear energy plays a vital role in addressing global energy challenges, including reducing carbon emissions, meeting rising energy demands, and transitioning to a more sustainable energy mix.

Conclusion:

The nuclear reactor, as a source of abundant and clean energy, has the potential to shape the future of energy production. Its development and responsible use hold the key to addressing pressing global energy and environmental challenges.

Chapter 20
The Spacecraft

Introduction:

The spacecraft represents the pinnacle of human engineering and innovation, allowing us to venture beyond Earth's confines and explore the vast expanse of space. In this chapter, we delve into the captivating story of the spacecraft and its role in exploring the final frontier.

Early Visions of Space Travel:

Humanity's fascination with space travel dates back centuries, with early visions of reaching the stars and exploring the cosmos. These dreams became a reality in the 20th century.

The Space Race:

The Cold War rivalry between the United States and the Soviet Union in the mid-20th century ignited the space race. The launch of Sputnik 1 by the Soviets in 1957 marked the beginning of human space exploration.

The Mercury and Apollo Programs:

NASA's Mercury and Apollo programs aimed at sending astronauts into space and, ultimately, to the Moon. In 1969, Apollo 11 achieved the historic Moon landing, marking a monumental milestone in human history.

Space Shuttles and the Era of Reusability:

The Space Shuttle program, which began in the 1980s, introduced reusable spacecraft, enabling routine access to space. Shuttles carried astronauts, conducted experiments, and deployed satellites.

Unmanned Spacecraft and Robotic Exploration:

Unmanned spacecraft, including probes, rovers, and orbiters, have played a pivotal role in exploring our solar system and beyond. Missions like Voyager, Hubble, and Mars rovers have provided invaluable data.

The International Space Station (ISS):

The construction and operation of the ISS, a joint effort involving multiple countries, have provided a platform for scientific research and international cooperation in space.

Commercial Spaceflight:

The 21st century has witnessed the rise of commercial spaceflight, with companies like SpaceX and Blue Origin competing to provide commercial access to space and reduce the cost of launches.

Challenges of Deep Space Exploration:

Deep space exploration presents immense challenges, including prolonged missions, radiation exposure, and the need for life support systems and sustainable resources.

The Search for Extra-terrestrial Life:

Spacecraft like the Mars rovers and telescopes like Kepler and TESS are dedicated to the search for extra-terrestrial life and habitable exoplanets.

The Future of Spacecraft:

The future of spacecraft holds exciting possibilities, including missions to Mars, asteroid mining, lunar bases, and the development of advanced propulsion technologies.

The Role of Spacecraft in Human Understanding:

Spacecraft have expanded our understanding of the universe, from the origins of our solar system to the nature of distant galaxies. They have also inspired generations with the wonder of space exploration.

Global Collaboration in Space Exploration:

International cooperation in space exploration fosters diplomacy and shared goals. Collaborative missions exemplify the potential for peaceful endeavours beyond Earth.

Conclusion:

The spacecraft, as a symbol of human ingenuity and curiosity, has propelled us to explore the final frontier. It has expanded our knowledge of the cosmos, pushed the boundaries of technology, and instilled a sense of wonder and inspiration.

Chapter 21
The Refrigerator

Introduction:

The refrigerator is a household appliance that has revolutionized the way we store and preserve perishable food items, extending their shelf life and reducing waste. In this chapter, we delve into the captivating story of the refrigerator and its role in preserving perishables.

Early Methods of Food Preservation:

Before the invention of the refrigerator, people relied on various methods to keep food from spoiling, including drying, salting, smoking, and fermentation. These methods had limitations and often altered the taste and texture of food.

The Need for Refrigeration:

The demand for a more effective and convenient way to preserve perishables grew with urbanization and the expansion of food distribution networks.

The Icebox: Precursor to the Refrigerator:

The first step toward refrigeration was the icebox, a wooden or metal container insulated with materials like straw or cork. Ice harvested from lakes and rivers was placed inside to keep food cool.

Invention of the Refrigerator:

The modern refrigerator was invented in the late 19th century. Several inventors and engineers contributed to its development, but the breakthrough came with the work of Carl von Linde and Albert Einstein on the principles of refrigeration.

How Refrigeration Works:

Refrigerators operate on the principle of vapor compression or absorption cycles. They remove heat from the interior, maintaining a temperature low enough to slow the growth of bacteria and fungi that cause food spoilage.

Advancements in Refrigeration Technology:

Over the years, refrigeration technology has seen numerous advancements, including the introduction of automatic defrosting, adjustable shelves, and energy-efficient compressors.

The Impact on Food Preservation:

Refrigeration has greatly improved food safety by slowing down the growth of pathogens and reducing the risk of foodborne illnesses. It has also extended the shelf life of perishable items.

Changing Food Habits:

The availability of refrigeration has influenced food habits and allowed for the storage of a wider variety of foods, including fresh fruits, vegetables, dairy products, and meats.

Commercial Refrigeration:

Refrigeration technology is essential in the food industry for preserving and storing bulk quantities of perishables, enabling global food distribution.

Challenges and Environmental Concerns:

Refrigeration technology relies on refrigerants, some of which have been linked to ozone depletion and climate change. Efforts are underway to develop more environmentally friendly alternatives.

Modern Refrigerators:

Today's refrigerators come equipped with features such as ice makers, water dispensers, digital temperature controls, and smart technology for remote monitoring and energy efficiency.

The Future of Refrigeration:

Research is ongoing to further improve refrigeration technology, with an emphasis on energy efficiency, sustainability, and reducing the environmental impact.

Conclusion:

The refrigerator, as a vital appliance for preserving perishables, has transformed the way we store and consume food. It has improved food safety, reduced waste, and allowed for greater culinary diversity.

Chapter 22
The Microwave Oven

Introduction:

The microwave oven stands as a culinary marvel that has revolutionized the way we cook and heat food, offering convenience, speed, and versatility in the kitchen. In this chapter, we delve into the captivating story of the microwave oven and its role as an instant culinary marvel.

Early Cooking Methods:

Before the microwave oven, cooking and heating food required traditional methods like stovetops, ovens, and open flames. These methods often took time and effort.

The Invention of the Microwave Oven:

The development of the microwave oven can be attributed to several inventors and scientists, with Percy Spencer often credited for the discovery. In 1945, he noticed that microwaves could heat food when a candy bar melted in his pocket near a magnetron, a component of radar equipment.

How Microwaves Cook:

Microwave ovens cook food by emitting microwave radiation that excites water molecules in food. This agitation generates heat, cooking the food quickly and evenly.

Early Microwave Ovens:

The first microwave ovens were large and expensive, primarily used in commercial settings like restaurants and ships. They were not initially considered practical for home use.

The Advent of Compact Home Microwaves:

In the 1960s, compact and affordable microwave ovens were introduced to the consumer market. They quickly gained popularity for their convenience and time-saving capabilities.

Microwaves and Culinary Versatility:

Microwave ovens are versatile appliances that can cook, reheat, defrost, and steam a wide variety of foods, from leftovers to vegetables and even desserts.

Convenience and Speed:

One of the microwave's primary advantages is speed. It can heat or cook food in a matter of minutes, making it ideal for busy households.

Popularity and Integration into Daily Life:

Microwave ovens have become commonplace in kitchens worldwide, changing the way we prepare and consume meals.

Frozen and Convenience Foods:

Microwave ovens have fuelled the growth of the frozen food industry, offering quick and easy meal options.

Challenges and Concerns:

Despite their convenience, microwave ovens have raised concerns about food safety and potential health risks associated with microwave radiation, although research has shown them to be safe when used correctly.

Modern Microwave Technology:

Advancements in microwave technology have led to features like sensor cooking, inverter technology, and pre-set cooking modes, enhancing cooking precision and convenience.

The Future of Microwave Ovens:

Future developments may focus on further energy efficiency, smart home integration, and improved cooking performance.

Conclusion:

The microwave oven, as an instant culinary marvel, has transformed the way we cook and eat. It has saved time, simplified meal preparation, and introduced new possibilities in the kitchen.

Chapter 23
Air Conditioning

Introduction:

Air conditioning, often taken for granted in our modern lives, is a remarkable invention that has profoundly shaped our comfort, productivity, and living conditions. In this chapter, we delve into the captivating story of air conditioning and its pivotal role in cooling the modern world.

Early Quest for Climate Control:

The desire for climate control dates back to ancient civilizations, with attempts to cool indoor spaces using methods like hand fans, water-soaked reeds, and clever architectural design.

The Birth of Mechanical Air Conditioning:

The advent of mechanical air conditioning can be attributed to Willis Haviland Carrier, who invented the first modern air conditioning system in 1902. Carrier's design aimed at controlling temperature and humidity for a printing plant.

The Principle of Air Conditioning:

Air conditioning systems work on the principles of refrigeration. They remove heat from indoor spaces and release it outside, creating a controlled and comfortable indoor environment.

Early Applications:

The initial applications of air conditioning were primarily industrial and commercial, such as cooling factories, movie theatres, and department stores. It revolutionized industries that required precise temperature and humidity control.

Expanding into Residential Use:

Air conditioning gradually found its way into residential homes, transforming the way people lived in hot and humid climates. It spurred population growth in regions like the American South and the Sun Belt.

Economic and Social Impact:

The widespread adoption of air conditioning had a profound economic and social impact. It increased productivity in workplaces, extended business hours, and influenced architectural design.

Air Conditioning and Medicine:

Air conditioning played a pivotal role in healthcare settings, enabling the control of indoor temperatures and humidity for patients' comfort and recovery.

Energy Efficiency and Environmental Concerns:

As air conditioning systems became more common, there were concerns about energy consumption and the environmental impact of refrigerants. Advances in technology have led to more energy-efficient and eco-friendly solutions.

The Global Spread of Air Conditioning:

Air conditioning has become a global phenomenon, with countries around the world adopting it as a standard feature in homes, offices, and vehicles.

Challenges of Extreme Heat:

Air conditioning has become essential in mitigating the health risks associated with extreme heatwaves, protecting vulnerable populations.

Innovation in Air Conditioning Technology:

Technological advancements have led to innovations such as ductless mini-split systems, smart thermostats, and energy-efficient cooling methods.

Future of Air Conditioning:

The future of air conditioning involves continued advancements in efficiency, sustainability, and the integration of smart technology for precise climate control.

Conclusion:

Air conditioning, as a key player in cooling the modern world, has reshaped our living conditions, increased our comfort, and had a profound impact on industries, architecture, and society as a whole.

Chapter 24
The Automobile

Introduction:

The automobile, a symbol of mobility and independence, has played a transformative role in modern society, reshaping transportation, commerce, and culture. In this chapter, we delve into the captivating story of the automobile and its role as a source of freedom on wheels.

Pioneers of the Automobile:

The development of the automobile can be attributed to inventors and engineers such as Karl Benz, Gottlieb Daimler, and Henry Ford. Their innovations laid the foundation for the automotive industry.

The Birth of the Motor Car:

In 1885, Karl Benz introduced the Motorwagen, considered the world's first true automobile. It was powered by an internal combustion engine and marked the beginning of the automotive era.

The Mass Production Revolution:

Henry Ford's introduction of the assembly line in 1913 revolutionized car manufacturing, making automobiles more affordable and accessible to the masses.

The Impact on Transportation:

The automobile transformed transportation, offering a level of convenience, speed, and personal mobility previously unimaginable. It led to the construction of roads and highways and the development of urban planning.

Automobile Culture and Identity:

Cars became symbols of personal identity and social status. They were associated with freedom, adventure, and the open road. Car culture thrived, with enthusiasts customizing and racing their vehicles.

The Role of the Automobile in World Wars:

During World War I and World War II, automobiles played crucial roles as military vehicles, ambulances, and transport for goods and troops.

The Rise of Automakers:

The 20th century saw the emergence of major automakers like General Motors, Chrysler, and Toyota, shaping the global automotive industry.

Environmental Concerns and Technological Innovations:

As the environmental impact of automobiles became a concern, the industry focused on innovations like catalytic converters, fuel efficiency, and hybrid and electric vehicles.

The Future of Automobiles:

The future of automobiles includes autonomous vehicles, electric and hydrogen-powered cars, and advancements in connectivity and safety features.

Automobiles in Pop Culture:

Automobiles have left an indelible mark on popular culture, featuring prominently in films, music, and literature.

Challenges and Opportunities:

The automotive industry faces challenges related to sustainability, urban congestion, and shifting consumer preferences. However, it also presents opportunities for innovation and growth.

Conclusion:

The automobile, as a symbol of freedom on wheels, has transformed the way we live, work, and travel. It has enabled personal mobility, shaped our culture, and driven technological progress.

Chapter 25
The Bicycle

Introduction:

The bicycle, a simple yet ingenious invention, has played a pivotal role in human transportation, recreation, and progress. In this chapter, we delve into the captivating story of the bicycle and its role as the pedals of progress.

Early Precursors:

The concept of using pedals and wheels for transportation dates back centuries, with early inventions like the draisine or hobby horse and the velocipede serving as precursors to the bicycle.

The Birth of the Bicycle:

In the early 19th century, Karl Drais, a German inventor, introduced the Laufmaschine or "running machine," often considered the first true bicycle. It featured a wooden frame and a front-wheel steering mechanism.

The Development of the Safety Bicycle:

The design of the safety bicycle, with two wheels of equal size and a chain drive, evolved in the late 19th century. This design made cycling more stable and accessible to a broader range of riders.

The Bicycle's Role in Women's Liberation:

The safety bicycle was instrumental in the late 19th-century women's liberation movement, providing women with newfound mobility and independence.

The Bicycle Boom:

The late 19th and early 20th centuries saw a bicycle boom, with cycling becoming a popular recreational activity and a means of transportation.

Bicycles in World Wars:

Bicycles played roles in both World War I and World II, serving as efficient means of transportation for troops, messengers, and reconnaissance.

Bicycles as a Mode of Transportation:

In many parts of the world, bicycles continue to be essential modes of transportation, providing an eco-friendly, efficient, and affordable means of getting around.

Cycling for Health and Fitness:

Cycling offers numerous health benefits, including cardiovascular exercise, improved mental well-being, and reduced environmental impact.

Bicycles and Environmental Sustainability:

Bicycles are celebrated for their role in reducing carbon emissions, alleviating traffic congestion, and promoting sustainable urban transportation.

Innovations in Bicycle Technology:

Advancements in materials, design, and technology have led to the development of various types of bicycles, including mountain bikes, road bikes, and electric bikes.

Bicycles in Competitive Sports:

Bicycles are integral to competitive sports like road cycling, mountain biking, and track cycling, where athletes push the limits of speed and endurance.

Challenges and Opportunities:

Challenges in promoting cycling include safety concerns, infrastructure development, and promoting cycling culture. Opportunities lie in further embracing bicycles as a sustainable urban transportation solution.

Conclusion:

The bicycle, as the pedals of progress, has empowered individuals, promoted health and fitness, and contributed to a more sustainable and interconnected world. Its legacy continues to evolve, shaping transportation, culture, and the way we live.

Chapter 26
The Sewing Machine

Introduction:

The sewing machine, a marvel of engineering simplicity, has revolutionized the textile and fashion industries, transforming the way we create clothing and textiles. In this chapter, we delve into the captivating story of the sewing machine and its role in stitching the fabric of progress.

Early Methods of Sewing:

Before the sewing machine, the production of clothing and textiles was a time-consuming and labour-intensive task. Hand sewing was the primary method, limiting the pace of textile production.

The Invention of the Sewing Machine:

The development of the sewing machine can be attributed to several inventors, with Elias Howe and Isaac Singer being notable figures in its history. In 1846, Elias Howe patented the first practical sewing machine, featuring a lockstitch mechanism.

How Sewing Machines Work:

Sewing machines work by using a needle and thread to create stitches in fabric. They automate the repetitive process of pushing the needle through the fabric, significantly increasing sewing speed and precision.

Revolutionizing the Textile Industry:

The sewing machine's introduction to the textile industry revolutionized production, enabling faster and more efficient manufacturing of clothing, shoes, and textiles.

The Role of Sewing Machines in Fashion:

Sewing machines allowed fashion designers to experiment with new styles and materials, leading to the rapid evolution of fashion trends.

Sewing Machines and Home Sewing:

Sewing machines eventually found their way into households, empowering individuals to create their clothing, repair textiles, and express their creativity.

Industrial Impact:

Sewing machines played a crucial role in the garment industry, making mass production of clothing possible. The sewing machine's influence extended to factories worldwide.

Advancements in Sewing Technology:

Over time, sewing machines evolved with features like automatic threading, zigzag stitching, and computerized controls, enhancing their versatility and ease of use.

The Sewing Machine's Impact on Employment:

While the sewing machine increased efficiency in textile production, it also had social and economic implications, impacting the livelihoods of skilled hand sewers.

Craftsmanship and Quality:

Despite the automation of sewing, craftsmanship and quality remain essential aspects of the textile and fashion industries.

Challenges and Opportunities:

Sewing machines continue to be essential tools, but they face challenges related to sustainability, ethical manufacturing, and the impact of automation on employment in the industry.

The Future of Sewing Technology:

Future innovations may focus on integrating technology, such as 3D printing and smart fabrics, into sewing machines, further expanding their capabilities.

Conclusion:

The sewing machine, as a vital instrument for stitching the fabric of progress, has transformed the textile and fashion industries, empowering individuals, shaping fashion trends, and contributing to economic development.

Chapter 27
The Locomotive

Introduction:

The locomotive, a symbol of industrial revolution and human ingenuity, has left an indelible mark on transportation, commerce, and the expansion of nations. In this chapter, we delve into the captivating story of the locomotive and its role as a powerful force on the rails of progress.

Early Transportation Challenges:

Before the locomotive, transportation relied on animal power, waterways, and human effort. The need for a more efficient and reliable mode of transportation became apparent as industrialization advanced.

The Birth of the Locomotive:

The development of the locomotive is attributed to several inventors, but George Stephenson is often credited with creating the first successful steam locomotive. In 1814, he built the "Blucher," a high-pressure steam engine that powered a locomotive.

Steam-Powered Revolution:

The locomotive harnessed the power of steam to drive its wheels, enabling it to pull heavy loads over long distances. This innovation revolutionized transportation and led to the rapid expansion of railways.

The First Successful Steam Locomotives:

In 1829, George Stephenson's "Rocket" won the Rainhill Trials, proving the viability of steam locomotives for practical use. The "Rocket" introduced design elements like multi-tubular boilers and a blast pipe that became standard in locomotive design.

Railway Expansion and Economic Growth:

The widespread adoption of locomotives and the construction of railways fuelled economic growth, connecting cities and regions and facilitating the movement of goods and people.

The Role of Railways in Nation Building:

Locomotives played a pivotal role in nation-building efforts, contributing to the unification and expansion of nations, including the United States and Canada.

Industrial Impact:

Locomotives and railways transformed industries like coal mining, steel production, and agriculture by providing efficient transportation for raw materials and finished products.

Innovation in Locomotive Technology:

Advancements in locomotive technology led to more powerful and efficient engines, streamlined designs, and safety features that improved reliability and safety.

The Golden Age of Rail Travel:

The 19th and early 20th centuries saw the emergence of luxurious passenger trains, offering unparalleled comfort and style for travellers.

The Decline of Steam Locomotives:

The mid-20th century brought the transition from steam to diesel and electric locomotives, marking the end of the steam era.

Challenges and Opportunities:

The railroad industry continues to face challenges related to infrastructure maintenance, sustainability, and competition from other modes of transportation. Opportunities for high-speed rail and improved connectivity remain on the horizon.

The Locomotive's Legacy:

The locomotive's legacy lives on in heritage railways, museums, and its enduring symbol of progress and human achievement.

Conclusion:

The locomotive, as a driving force on the rails of progress, has transformed transportation, commerce, and nations themselves. It has left an indomitable mark on history, symbolizing the power of innovation and the ability of humanity to overcome geographical barriers.

Chapter 28
The Camera

Introduction:

The camera, a device designed to capture and preserve images, has forever altered our relationship with the visual world, enabling us to freeze moments in time and relive them at our leisure. In this chapter, we delve into the captivating story of the camera and its role in capturing moments in time.

Early Attempts at Image Capture:

The human desire to capture images dates back centuries. Early attempts involved techniques like camera obscura, which projected an inverted image onto a surface.

The Invention of the Camera Obscura:

The camera obscura, a precursor to the modern camera, evolved over time. It consisted of a darkened room or box with a small hole or lens through which external scenes were projected onto a surface inside.

The Birth of Photography:

The development of photography can be attributed to various inventors and scientists. However, Louis Daguerre and William Henry Fox Talbot are often credited with pioneering early photographic processes.

Daguerreotype and Calotype:

Louis Daguerre's daguerreotype and Fox Talbot's calotype were among the first practical photographic processes, capturing images on treated metal plates and paper, respectively.

Evolution of Photographic Technology:

The evolution of photographic technology led to improvements in image quality, exposure times, and portability. Wet collodion plates and glass negatives were among the early advancements.

George Eastman and the Kodak Camera:

In 1888, George Eastman introduced the Kodak camera, a portable device that made photography accessible to a broader audience. It came pre-loaded with film and was marketed with the famous slogan, "You press the button, we do the rest."

The Role of Photography in Society:

Photography revolutionized journalism, art, science, and personal documentation. It played a significant role in capturing historical events, shaping public opinion, and preserving memories.

Instant Photography:

Edwin Land's Polaroid camera, introduced in the mid-20th century, allowed users to obtain instant prints, eliminating the need for film processing.

Digital Revolution:

The transition from film to digital photography marked a seismic shift in the industry. Digital cameras offered immediate feedback, ease of sharing, and virtually limitless storage.

Smartphone Cameras:

The integration of high-quality cameras into smartphones has made photography ubiquitous, allowing individuals to capture and share moments instantly.

Photography as Art:

Photography is celebrated as a powerful medium for artistic expression, with photographers exploring various styles, techniques, and themes.

Challenges and Opportunities:

Photography faces challenges related to privacy, digital manipulation, and the proliferation of images. However, it also presents opportunities for creative expression, storytelling, and documenting the world.

The Future of Photography:

The future of photography holds promise in emerging technologies like computational photography, virtual reality, and augmented reality, pushing the boundaries of visual storytelling.

Conclusion:

The camera, as a tool for capturing moments in time, has transformed how we perceive and interact with the world. It has become an integral part of our daily lives, allowing us to document our experiences, share stories, and preserve memories.

Chapter 29
The X-ray Machine

Introduction:

The X-ray machine, a ground-breaking invention in the realm of medical imaging, has revolutionized healthcare by allowing us to peer inside the human body without invasive procedures. In this chapter, we delve into the captivating story of the X-ray machine and its role in enabling us to see the inner workings of the human body.

Early Quest for Medical Insight:

Before the X-ray machine, physicians had limited tools to diagnose and understand internal injuries or diseases. They relied on clinical observations, palpation, and exploratory surgery.

Discovery of X-rays:

The discovery of X-rays is credited to Wilhelm Conrad Roentgen, a German physicist. In 1895, while experimenting with cathode rays, Roentgen noticed that a fluorescent screen in his lab emitted light even when shielded from the cathode rays, indicating the presence of a previously unknown form of radiation.

Unveiling the Invisible:

Roentgen's discovery revealed the existence of X-rays, a form of electromagnetic radiation with the ability to pass through soft tissues while being absorbed by denser materials like bones.

The First X-ray Image:

Roentgen's first X-ray image was of his wife's hand, which clearly displayed her bones and wedding ring. This ground-breaking image marked the beginning of medical radiography.

Medical Applications:

The X-ray machine's introduction to medicine revolutionized diagnosis and treatment. Physicians could now visualize bone fractures, locate foreign objects, and detect internal diseases.

Rapid Advancements:

The field of radiology advanced rapidly, with improvements in X-ray machine design, image quality, and safety. Lead aprons and collimators helped protect patients and healthcare providers from excessive radiation exposure.

X-ray Technology During World Wars:

X-ray technology played a vital role in military medicine during World War I and World War II, enabling the diagnosis of injuries and the planning of surgical procedures.

Fluoroscopy:

Fluoroscopy, a real-time X-ray imaging technique, allowed for dynamic visualization of internal structures during medical procedures such as barium swallow tests and angiography.

Computed Tomography (CT):

The development of CT scanners in the mid-20th century represented a significant advancement, providing detailed cross-sectional images of the body.

Magnetic Resonance Imaging (MRI):

MRI technology, which emerged in the 1980s, uses magnetic fields and radio waves to create detailed images of soft tissues, offering a non-invasive alternative to X-rays.

Positron Emission Tomography (PET) and Single Photon Emission Computed Tomography (SPECT):

PET and SPECT scans use radioactive tracers to create images that reveal metabolic activity and disease processes within the body.

Challenges and Safety Concerns:

While immensely valuable, X-ray technology is not without risks. Prolonged or excessive exposure to X-rays can damage DNA and increase the risk of cancer, prompting efforts to minimize radiation doses.

The Future of Medical Imaging:

Advances in medical imaging continue, with innovations like 3D imaging, molecular imaging, and artificial intelligence-assisted diagnostics shaping the future of healthcare.

Conclusion:

The X-ray machine, as a tool for peering inside the human body, has transformed medical diagnosis and treatment. It has saved countless lives, guided surgical procedures, and provided invaluable insights into the mysteries of human anatomy and pathology.

Chapter 30
The Vaccination

Introduction:

Vaccination, a medical marvel, has been a cornerstone of public health for centuries, preventing deadly diseases and saving countless lives. In this chapter, we delve into the captivating story of vaccination and its pivotal role in safeguarding humanity against infectious threats.

Early Struggles with Infectious Diseases:

Throughout history, infectious diseases like smallpox, measles, and polio have ravaged populations, causing immense suffering and death. Communities were often powerless in the face of these invisible killers.

The Concept of Immunization:

The concept of immunization, or inducing immunity against diseases, has ancient roots. Practices like variolation in China and India involved exposing individuals to small amounts of infectious material to confer protection.

The Smallpox Vaccine:

The smallpox vaccine, a milestone in medical history, is often attributed to Edward Jenner, an English physician. In 1796, he demonstrated that exposure to cowpox could protect against smallpox.

Eradication of Smallpox:

The success of the smallpox vaccine paved the way for the global eradication of smallpox, a remarkable achievement announced by the World Health Organization (WHO) in 1980.

The Advent of Modern Vaccines:

The development of vaccines against diseases like cholera, tetanus, and diphtheria followed, ushering in the era of modern vaccination.

Jonas Salk and the Polio Vaccine:

Jonas Salk's development of the inactivated polio vaccine in the 1950s marked a major breakthrough in preventing a disease that had paralyzed millions.

Mass Vaccination Campaigns:

Mass vaccination campaigns, often supported by governments and international organizations, have led to the control and elimination of diseases like measles, rubella, and hepatitis.

Vaccination's Role in Herd Immunity:

Vaccination not only protects individuals but also contributes to herd immunity, reducing the transmission of diseases within communities and protecting vulnerable populations.

Challenges and Vaccine Hesitancy:

Vaccine hesitancy, fuelled by misinformation and mistrust, poses challenges to vaccination efforts and public health. Addressing these concerns is crucial to maintaining high vaccination rates.

The COVID-19 Pandemic:

The COVID-19 pandemic underscored the importance of vaccines in controlling the spread of infectious diseases. The rapid development and distribution of COVID-19 vaccines have been a monumental achievement.

Future Frontiers in Vaccinology:

Advancements in vaccine technology, including mRNA vaccines and novel adjuvants, hold promise for tackling emerging infectious diseases and improving vaccine efficacy.

Equitable Vaccine Access:

Ensuring equitable access to vaccines worldwide is a global priority, with initiatives like COVAX working to distribute vaccines to low- and middle-income countries.

Conclusion:

Vaccination, as a powerful tool in preventing deadly diseases, has transformed public health and saved countless lives. It represents the triumph of science, collaboration, and the human spirit in the face of infectious threats.

Chapter 31
The GPS

Introduction:

The Global Positioning System (GPS), a marvel of modern technology, has revolutionized the way we navigate and find our way with pinpoint precision. In this chapter, we delve into the captivating story of the GPS and its pivotal role in guiding us accurately across the globe.

Early Navigation Challenges:

Navigating across vast distances, whether on land, sea, or in the air, presented significant challenges throughout history. Reliance on celestial objects, landmarks, and compasses had limitations.

The Emergence of GPS:

The development of the GPS can be attributed to the U.S. Department of Defence, which launched the first satellite, "Navstar 1," in 1978. The system was initially intended for military use.

How GPS Works:

The GPS consists of a network of satellites orbiting the Earth. These satellites transmit signals to GPS receivers on the ground, allowing them to calculate their precise location based on the time it takes for signals to reach them.

Accuracy and Precision:

One of the remarkable aspects of GPS is its high level of accuracy and precision. It can determine locations with accuracies down to a few meters or even centimetres, depending on the technology used.

Civilian and Commercial Use:

While developed for military purposes, the U.S. government opened GPS for civilian and commercial use in the 1980s. This decision transformed navigation in countless industries, including transportation, agriculture, and surveying.

GPS in Daily Life:

GPS has become an integral part of modern life. It powers navigation apps in smartphones, assists in finding locations, tracks fitness activities, and guides autonomous vehicles.

Aviation and Maritime Navigation:

GPS has drastically improved navigation in aviation and maritime industries, ensuring safer and more efficient travel.

Search and Rescue:

GPS has been a critical tool in search and rescue operations, helping locate lost or distressed individuals quickly and accurately.

Precision Agriculture:

In agriculture, GPS-guided tractors and equipment allow for precise planting, fertilizing, and harvesting, improving crop yields and reducing environmental impact.

Geospatial Sciences:

GPS has revolutionized geospatial sciences, enabling cartographers, surveyors, and scientists to create detailed maps and monitor changes in the Earth's surface.

Challenges and Future Developments:

Challenges related to signal accuracy, interference, and cybersecurity persist. However, the future of GPS holds innovations such as the use of additional satellite constellations and advanced receivers.

International Navigation Systems:

GPS is one of several global navigation satellite systems (GNSS), including GLONASS (Russia), Galileo (EU), and BeiDou (China), further enhancing global coverage and accuracy.

Conclusion:

The GPS, as a tool for navigating with precision, has transformed how we move, work, and explore the world. Its impact extends across industries and has made the world more accessible and interconnected.

Chapter 32
The Satellite

Introduction:

Satellites, the unsung heroes of modern technology, have transformed our world by providing a view from above, enabling a wide range of applications, from communication and weather forecasting to navigation and Earth observation. In this chapter, we delve into the captivating story of satellites and their pivotal role as "Eyes in the Sky."

Early Dreams of Space Observation:

The idea of observing Earth from space has intrigued scientists and visionaries for centuries. However, it wasn't until the mid-20th century that technology caught up with this dream.

The Dawn of the Space Age:

The launch of the Soviet satellite Sputnik 1 in 1957 marked the beginning of the space age. It was the first artificial satellite to orbit Earth, sparking a space race between the United States and the Soviet Union.

The Birth of Communication Satellites:

The launch of the first communication satellite, Echo 1, in 1960 paved the way for global telecommunications. Satellites in geostationary orbit could relay signals across continents, connecting people like never before.

Weather and Earth Observation:

Satellites equipped with sensors and cameras have revolutionized weather forecasting, allowing us to monitor storms, track climate patterns, and understand Earth's changing environment.

The Global Positioning System (GPS):

GPS, a constellation of satellites, provides accurate navigation and positioning information to users worldwide. It has applications in aviation, maritime, transportation, and everyday life.

Earth Imaging and Remote Sensing:

Satellites equipped with high-resolution cameras and sensors capture detailed images of Earth's surface, aiding in agriculture, forestry, disaster management, and urban planning.

Space Exploration:

Satellites have played pivotal roles in space exploration missions, enabling communication with spacecraft, mapping other celestial bodies, and monitoring the cosmos.

International Collaboration:

The International Telecommunication Union (ITU) and international agreements ensure equitable access to orbital slots and radio frequencies for satellite communication.

Miniaturization and CubeSats:

Advancements in miniaturization have led to the development of CubeSats, small satellites with diverse applications, democratizing access to space for research and experimentation.

Challenges and Sustainability:

Space debris and congestion in Earth's orbit pose challenges to satellite operations and sustainability. Efforts are underway to address these concerns.

Space Entrepreneurs and Commercial Space Industry:

Private companies like SpaceX, Blue Origin, and OneWeb are driving innovation in satellite technology and space exploration, opening new frontiers in commercial space ventures.

Future Frontiers:

The future of satellites includes mega-constellations for global internet coverage, space-based solar power, and Earth-observing satellites with advanced capabilities.

Conclusion:

Satellites, as "Eyes in the Sky," have transformed how we see, understand, and interact with our world and the universe beyond. Their impact spans communication, navigation, observation, and exploration, making the impossible possible.

Chapter 33
The Microscope

Introduction:

The microscope, a marvel of optics and scientific ingenuity, has allowed us to explore and understand the hidden universe of the microscopic world. In this chapter, we delve into the captivating story of the microscope and its pivotal role in revealing the mysteries of the small-scale world.

Early Curiosity About the Microscopic World:

The fascination with the small and unseen traces back to ancient times when early scientists and philosophers speculated about the existence of tiny, invisible entities.

The Birth of Microscopy:

The development of the microscope as we know it can be attributed to the Dutch spectacle maker Zacharias Janssen, who, in the late 16th century, created the first compound microscope by combining lenses in a tube.

Antonie van Leeuwenhoek's Discoveries:

The Dutch scientist Antonie van Leeuwenhoek is often referred to as the "Father of Microbiology" for his ground-breaking work with microscopes. In the 17th century, he used his handcrafted single-lens microscopes to observe microorganisms for the first time, describing bacteria, protozoa, and other microscopic life forms.

The Microscope's Role in Advancing Science:

Microscopes revolutionized biology, enabling scientists to explore the intricacies of cells, tissues, and organisms. They contributed to the development of the cell theory and expanded our understanding of life at its fundamental level.

The Electron Microscope:

In the 20th century, the invention of the electron microscope by Ernst Ruska and Max Knoll dramatically improved resolution and allowed scientists to see even smaller structures. Transmission electron microscopes (TEM) and scanning electron microscopes (SEM) have since become essential tools in various scientific disciplines.

Applications in Medicine:

Microscopes have played a critical role in medical diagnostics and research, allowing physicians to examine blood cells, bacteria, and tissue samples, leading to advances in pathology and microbiology.

Materials Science and Nanotechnology:

Microscopes have been instrumental in materials science and nanotechnology, enabling scientists to study the properties and structures of materials at the nanoscale.

Microscopy in Industry:

In industry, microscopes are used for quality control, failure analysis, and research and development, contributing to advances in manufacturing and technology.

Forensics and Criminal Investigations:

Microscopes have been crucial in forensic science, helping investigators analyse trace evidence, identify counterfeit materials, and solve crimes.

Modern Microscopy Techniques:

Advancements in microscopy include confocal microscopy, fluorescence microscopy, and super-resolution microscopy, which have expanded our ability to study biological processes and structures with unprecedented detail.

Challenges and Future Directions:

Challenges in microscopy include achieving even higher resolutions, reducing sample damage in electron microscopy, and developing portable and affordable microscope technologies.

Conclusion:

The microscope, as a tool for unveiling the microscopic world, has transformed how we perceive life, materials, and the natural world. It has unlocked mysteries that were once hidden from our view, expanding our understanding of the intricate and fascinating realms of the small-scale universe.

Chapter 34
The ATM

Introduction:

The Automated Teller Machine (ATM), a financial innovation that revolutionized banking and access to cash, has become an indispensable part of modern life. In this chapter, we delve into the captivating story of the ATM and its pivotal role in providing instant access to cash.

Early Banking and Cash Withdrawal:

Before the advent of ATMs, banking and cash withdrawal were often confined to banking hours and in-person visits to bank branches. This limited access posed significant challenges for individuals needing funds outside of regular business hours.

The Birth of the ATM:

The concept of the ATM can be attributed to John Shepherd-Barron, a Scottish inventor who, in the 1960s, developed the idea of a machine that could dispense cash to customers using a secure system of Personal Identification Numbers (PINs).

The First ATM Installation:

The world's first ATM was installed by Barclays Bank in London in 1967. It was a significant leap in banking technology, allowing customers to access their accounts and withdraw cash at any time, day or night.

ATM Evolution:

ATM technology continued to evolve, with improvements in security, accessibility, and functionality. Magnetic stripe cards, which replaced paper vouchers, and colour screens were among the innovations.

Global ATM Networks:

The concept of the ATM quickly spread beyond the UK, leading to the development of global ATM networks that allow users to access cash from their accounts, irrespective of their location.

Enhanced Banking Services:

ATMs have expanded their services beyond cash withdrawal, enabling users to deposit funds, transfer money, check balances, and even purchase stamps, lottery tickets, and event tickets.

Accessibility and Inclusivity:

ATMs have played a vital role in improving banking accessibility, making financial services available to individuals in remote areas and enhancing financial inclusion.

ATMs in Emergencies:

During natural disasters and emergencies, ATMs provide essential access to cash for individuals and communities in need.

Security Measures:

Security features such as PINs, encryption, and biometric authentication have been implemented to protect ATM users and prevent fraud.

The Digital Era:

ATMs have adapted to the digital era, with the integration of contactless card technology, mobile banking apps, and cardless ATM transactions.

Challenges and Future Directions:

ATMs face challenges related to cybersecurity and emerging payment technologies. However, their role in providing convenient cash access remains essential.

Conclusion:

The ATM, as a means of providing instant access to cash, has transformed the way we conduct financial transactions and manage our finances. It has made banking more convenient, accessible, and inclusive, impacting individuals and businesses alike.

Chapter 35
The Laser

Introduction:

The laser, a revolutionary invention, has transformed various fields of science, technology, and industry with its ability to produce highly focused and coherent beams of light. In this chapter, we delve into the captivating story of the laser and its pivotal role in enabling precision light amplification.

Early Light Amplification Concepts:

The concept of light amplification can be traced back to the late 19th century when scientists explored the possibility of amplifying light through various means, such as stimulated emission.

The Birth of the Laser:

The term "laser" stands for "Light Amplification by Stimulated Emission of Radiation." The development of the laser can be credited to Theodore H. Maiman, who, in 1960, built the first working laser using a synthetic ruby crystal.

Principles of Laser Operation:

Laser operation is based on the principles of stimulated emission, where atoms or molecules emit photons of light when stimulated by external photons of the same wavelength. This process leads to the generation of a coherent and focused beam of light.

Types of Lasers:

Lasers come in various types, including gas lasers, solid-state lasers, semiconductor lasers, and fibre lasers, each with unique properties and applications.

Industrial and Manufacturing Applications:

Lasers have revolutionized industrial processes such as cutting, welding, and engraving by providing precise and controlled energy delivery. They are essential tools in manufacturing industries.

Medical and Surgical Applications:

In medicine, lasers are used for various purposes, including surgery, dermatology, dentistry, and vision correction. Laser technology allows for minimally invasive procedures with reduced scarring and quicker recovery times.

Communication and Information Technology:

Lasers play a crucial role in communication through optical fibres, enabling high-speed data transmission over long distances. They are also used in barcode scanners and laser printers.

Scientific Research and Exploration:

Lasers are indispensable tools in scientific research, from studying atomic and molecular structures to conducting experiments in physics and chemistry. They are used in the creation of super-intense, short-duration laser pulses for studying ultrafast phenomena.

Entertainment and Imaging:

Lasers are used in entertainment, such as laser light shows and laser projectors, creating stunning visual effects. They are also utilized in various imaging techniques, including laser scanning microscopy.

Laser Interferometry:

Laser interferometry is a precise measurement technique used in fields like astronomy and gravitational wave detection, allowing scientists to detect minute changes in distances.

Challenges and Safety Considerations:

Lasers can pose safety hazards due to their high intensity. Regulations and safety measures are in place to mitigate risks associated with laser use.

Future Developments:

Ongoing research and development in laser technology aim to create even more powerful and efficient lasers, opening up new possibilities in science, technology, and medicine.

Conclusion:

The laser, as a tool for precision light amplification, has transformed multiple facets of our lives, from manufacturing and medicine to communication and scientific exploration. Its ability to produce focused and coherent light has revolutionized numerous industries, pushing the boundaries of human capability and understanding.

Chapter 36
The Robot

Introduction:

The robot, a remarkable creation of engineering and artificial intelligence, has revolutionized industries, labour, and even our daily lives with its ability to automate tasks and perform functions autonomously. In this chapter, we delve into the captivating story of the robot and its pivotal role in automating industry and beyond.

Early Automaton Concepts:

The concept of automating tasks dates back centuries, with early automata, mechanical devices designed to mimic human or animal movements, serving as precursors to modern robots.

The Birth of Robotics:

The term "robot" was coined by Czech playwright Karel Čapek in his 1920 play, "R.U.R." (Rossum's Universal Robots). The play introduced the idea of artificial, humanoid workers.

Early Industrial Robots:

The first industrial robots, developed in the 1950s and 1960s, were large, stationary machines used primarily in manufacturing for tasks like welding and assembly.

Robotics in Manufacturing:

Robots revolutionized manufacturing by increasing efficiency, precision, and speed. They could perform repetitive and dangerous tasks with consistent quality.

Types of Robots:

Robots come in various types, including industrial robots, service robots, medical robots, and autonomous vehicles, each designed for specific applications.

Service Robots:

Service robots are designed to assist humans in various roles, such as healthcare, agriculture, and cleaning. Examples include surgical robots, robotic vacuums, and delivery drones.

Medical Robotics:

Robotic systems have transformed healthcare, assisting surgeons in minimally invasive procedures, providing rehabilitation therapy, and even delivering medications in hospitals.

Robots in Space Exploration:

Robots, including rovers like Curiosity and Perseverance, have played a crucial role in space exploration, allowing scientists to explore distant planets and gather valuable data.

Consumer Robotics:

Consumer robotics, like home assistant robots and educational robots, have become increasingly popular, making technology more accessible to everyday users.

Artificial Intelligence and Machine Learning:

Advancements in artificial intelligence and machine learning have empowered robots with greater autonomy, adaptability, and decision-making capabilities.

Challenges and Ethical Considerations:

The rise of robotics has brought challenges related to job displacement, security, privacy, and ethical concerns surrounding autonomous weapons and AI ethics.

Future Directions:

The future of robotics holds promises such as human-robot collaboration, swarm robotics, and the integration of robots into smart cities and infrastructure.

Conclusion:

The robot, as a tool for automating industry and various aspects of human life, has transformed how we work, explore, and interact with the world. Its impact extends across industries, from manufacturing and healthcare to space exploration and everyday convenience.

Chapter 37
The Assembly Line

Introduction:

The assembly line, a ground-breaking innovation in manufacturing, has redefined the way products are produced by streamlining processes, increasing efficiency, and reducing costs. In this chapter, we delve into the captivating story of the assembly line and its pivotal role in revolutionizing production.

Early Manufacturing Challenges:

Before the assembly line, manufacturing was often a slow, labour-intensive process with artisans or workers crafting products one at a time, resulting in limited production output and higher costs.

The Birth of the Assembly Line:

The concept of the assembly line can be traced back to the late 18th century, but it was Henry Ford, the American automotive pioneer, who revolutionized manufacturing with the introduction of the moving assembly line in 1913.

The Ford Model T:

Henry Ford's implementation of the assembly line transformed the automobile industry. By breaking down the production of the Model T into discrete, standardized tasks, he reduced the time it took to assemble a car from 12 hours to just 93 minutes.

Key Principles of the Assembly Line:

The assembly line involves the division of labour, with each worker responsible for a specific task. Components and parts are moved along a conveyor system, allowing for continuous production and minimal handling.

Mass Production:

The assembly line enabled mass production, making products more affordable and accessible to a broader range of consumers. It also reduced manufacturing costs, contributing to increased profits.

Expansion to Other Industries:

The success of the assembly line in the automotive industry led to its adoption in various sectors, including electronics, aerospace, and consumer goods.

Impact on the Workforce:

While the assembly line increased efficiency, it also raised concerns about worker monotony and repetitive tasks. Labour unions and workplace regulations emerged to address these issues.

Automation and Robotics:

In modern manufacturing, automation and robotics have further enhanced the assembly line, allowing for even greater precision, speed, and flexibility.

Lean Manufacturing:

Lean manufacturing principles, inspired by the assembly line, focus on eliminating waste, improving efficiency, and maximizing value in production processes.

Challenges and Sustainability:

The assembly line, while efficient, has faced criticism for its environmental impact and the disposable culture it can promote. Efforts are underway to make production more sustainable.

Future Directions:

Advancements in technology, including 3D printing and smart manufacturing, continue to shape the future of production, offering new possibilities for customization and efficiency.

Conclusion:

The assembly line, as a tool for efficiency in production, has transformed how goods are manufactured, making products more accessible, affordable, and widely available. Its impact extends beyond manufacturing, influencing business practices and consumer expectations.

Chapter 38
The Solar Cell

Introduction:

The solar cell, a remarkable invention in renewable energy technology, has revolutionized the way we generate electricity by harnessing the abundant power of the sun. In this chapter, we delve into the captivating story of the solar cell and its pivotal role in converting sunlight into clean and sustainable energy.

Early Interest in Solar Energy:

The utilization of solar energy traces back to ancient civilizations that used magnifying glasses to concentrate sunlight for various purposes. However, it was in the 19th century that scientists began exploring ways to convert solar energy into electricity.

The Birth of the Solar Cell:

The modern solar cell, also known as a photovoltaic cell, was developed in the mid-20th century by Bell Labs researchers. In 1954, they created the first practical solar cell using silicon, which demonstrated the conversion of sunlight into electricity.

Photovoltaic Effect:

The operation of solar cells is based on the photovoltaic effect, where photons from sunlight strike semiconductor materials, generating electron-hole pairs and producing an electric current.

Types of Solar Cells:

Solar cells come in various types, including monocrystalline, polycrystalline, and thin-film, each with its unique characteristics and applications.

Solar Power Advancements:

Advancements in solar technology have led to increased efficiency, durability, and affordability of solar panels, making them a viable source of renewable energy.

Grid-Connected and Off-Grid Systems:

Solar power systems can be grid-connected, allowing excess electricity to be fed back into the grid, or off-grid, providing energy in remote areas or during power outages.

Applications in Space:

Solar cells are widely used in space exploration and satellites, where they provide a reliable source of power in the harsh conditions of space.

Residential and Commercial Solar Installations:

The installation of solar panels on rooftops and in commercial buildings has become increasingly popular, allowing individuals and businesses to generate their electricity and reduce their carbon footprint.

Solar Farms:

Large-scale solar farms, equipped with thousands of solar panels, contribute significantly to clean energy generation and the reduction of greenhouse gas emissions.

Environmental Benefits:

Solar power is a clean and renewable energy source that reduces greenhouse gas emissions, air pollution, and dependence on fossil fuels.

Challenges and Energy Storage:

Challenges in solar energy include intermittency (solar power generation is dependent on sunlight) and the need for effective energy storage solutions like batteries.

Future Directions:

The future of solar technology includes innovations in materials, improved energy storage, and integration with smart grids for efficient energy management.

Conclusion:

The solar cell, as a tool for harnessing the sun's energy, has transformed the energy landscape, offering a sustainable and eco-friendly alternative to fossil fuels. Its impact extends across industries, from residential and commercial power generation to space exploration.

Chapter 39
The Wind Turbine

Introduction:

The wind turbine, a pioneering invention in renewable energy technology, has revolutionized the way we generate electricity by harnessing the power of the wind. In this chapter, we delve into the captivating story of the wind turbine and its pivotal role in converting wind energy into clean and sustainable power.

Ancient Wind Power:

The utilization of wind energy has ancient roots, with civilizations using wind power for tasks like milling grain and pumping water. However, it was in the late 19th and early 20th centuries that wind turbines for electricity generation began to emerge.

The Birth of Modern Wind Turbines:

The modern wind turbine owes much of its development to inventors like Charles Brush and Albert Betz, who made significant contributions to wind turbine design and efficiency in the early 20th century.

Types of Wind Turbines:

Wind turbines come in various types, including horizontal-axis and vertical-axis turbines, each with its unique design and applications.

The Horizontal-Axis Wind Turbine:

The most common type of wind turbine, the horizontal-axis wind turbine, features blades that rotate around a horizontal axis. It has become the standard for large-scale wind farms.

Wind Farms and Power Generation:

Wind farms, consisting of multiple wind turbines strategically placed in areas with strong and consistent winds, have become a major source of renewable energy, capable of producing electricity for entire communities.

Offshore Wind Energy:

Offshore wind farms, located in bodies of water, take advantage of strong coastal winds. They offer the potential for even greater energy generation and are less visible on land.

Small-Scale and Residential Wind Turbines:

Small-scale wind turbines are suitable for residential and remote applications, providing clean power to individual homes and businesses.

Advancements in Wind Turbine Technology:

Advancements in materials, aerodynamics, and control systems have increased the efficiency and reliability of wind turbines.

Environmental Benefits:

Wind power is a clean and renewable energy source that reduces greenhouse gas emissions, air pollution, and dependence on fossil fuels.

Challenges and Grid Integration:

Challenges in wind energy include intermittency (wind power generation depends on wind speed) and the need for efficient grid integration and energy storage solutions.

Energy Storage and Hybrid Systems:

Energy storage solutions like batteries, combined with wind turbines, can provide reliable power even when the wind is not blowing.

Future Directions:

The future of wind energy includes innovations in turbine design, grid integration, and the use of artificial intelligence to optimize wind farm performance.

Conclusion:

The wind turbine, as a tool for harvesting wind energy, has transformed the energy landscape, offering a sustainable and eco-friendly alternative to fossil fuels. Its impact extends across industries, from large-scale wind farms to individual homes and businesses.

Chapter 40
The Arch

Introduction:

The arch, an ancient architectural marvel, has played a pivotal role in shaping the world's built environment. This architectural element has not only served as a functional and structural solution but has also showcased the ingenuity and artistry of human civilizations. In this chapter, we delve into the captivating story of the arch and its significant role in building marvels of engineering.

The Origins of the Arch:

The concept of the arch can be traced back to ancient civilizations, including the Mesopotamians and Egyptians, who used rudimentary forms of the arch in their architectural designs. However, it was the Romans who perfected the arch, contributing to its widespread use.

The Roman Arch:

Roman architects and engineers utilized the arch extensively in their constructions, from aqueducts and bridges to amphitheatres and monumental structures like the Colosseum. The arch's semi-circular shape distributed weight efficiently, allowing for larger and more ambitious architectural projects.

The Keystone:

A defining feature of the arch is the keystone, the central stone that locks the arch's elements in place. This unique feature distributes the arch's load, providing stability and strength.

Gothic Arch and Cathedrals:

During the Gothic period, architects employed pointed arches to create soaring cathedrals with intricate stained glass windows. These structures showcased not only structural innovation but also artistic expression.

The Triumphal Arch:

Triumphal arches were built by various ancient civilizations to commemorate military victories and leaders. The Arch of Titus in Rome and the Arc de Triomphe in Paris are iconic examples.

Advancements in Arch Technology:

As architecture evolved, so did arch technology. Innovations like the flying buttress and ribbed vaults allowed for even greater architectural heights and grandeur.

Modern Applications:

While arches are often associated with historical and monumental structures, they continue to find applications in modern architecture and engineering. The Sydney Opera House, with its shell-like arches, is a notable example.

Structural Significance:

The arch's strength lies in its ability to distribute loads and forces uniformly, making it an ideal choice for bridges, tunnels, and even underground construction.

Artistic Expression:

Beyond its structural prowess, the arch is a symbol of artistic expression and cultural identity, as seen in structures like the Gateway Arch in St. Louis, Missouri.

Challenges and Innovations:

Modern architects and engineers face challenges in designing arch-based structures that meet contemporary demands for sustainability, accessibility, and aesthetics. Innovations in materials and construction techniques continue to push the boundaries of arch design.

Conclusion:

The arch, as a marvel of engineering and artistry, has left an indelible mark on human history and the built environment. Its enduring appeal lies in its timeless beauty, structural integrity, and adaptability to various architectural styles and functions.

A Celebration

In closing, we have embarked on a remarkable journey through the annals of human ingenuity and innovation, exploring inventions that have shaped our world and enriched our lives. From the simplest tools of ancient civilizations to the cutting-edge technologies of the present day, these inventions have not only defined eras but have also transcended time, becoming an integral part of our collective human experience.

As we reflect on the transformative power of these inventions, we must recognize that our journey is far from over. In the grand tapestry of human history, we stand at a pivotal moment where the pace of innovation continues to accelerate. The same spirit of curiosity and discovery that led to the creation of the wheel, the printing press, the computer, and countless other inventions still burns brightly within us.

The future holds the promise of even more astonishing breakthroughs, ones that will undoubtedly reshape our world in profound ways. From advances in artificial intelligence and renewable energy to space exploration and medical discoveries, the horizon of human potential knows no bounds.

Inventions are not just artifacts of the past; they are the building blocks of our future. They are the tools we use to navigate the challenges and opportunities of our time. They are the bridges that connect us across generations and cultures, allowing us to share knowledge and dreams.

So, let us celebrate the inventors, the innovators, and the visionaries who have propelled us forward. Let us embrace the endless possibilities that lie ahead. As we turn the final page of this book, let us remember that the story of invention is an ever-unfolding narrative—one that we each have a role in shaping.

Inventions are not merely creations of the mind; they are reflections of the human spirit's unyielding desire to understand, create, and progress.

They remind us that our capacity for innovation is limitless, and our journey to explore the frontiers of knowledge and possibility is boundless.

With this knowledge in our hearts and a sense of wonder in our souls, we move forward, ever eager to write the next chapter of human invention. Inventions are the keys that unlock the doors to a brighter, more remarkable future—a future where the limits of what we can achieve are defined only by the boundless horizons of our imagination.

And so, as we conclude this exploration of the world's greatest inventions, we invite you to carry the torch of innovation forward, to dream boldly, and to shape the world for generations to come. For in the world of invention, there are no final chapters—only new beginnings, waiting to be written.

Don't miss out!

Visit the website below and you can sign up to receive emails whenever PA BOOKS publishes a new book. There's no charge and no obligation.

https://books2read.com/r/B-A-STTAB-EGXOC

BOOKS 2 READ

Connecting independent readers to independent writers.

Also by PA BOOKS

Hogan's Key
Kimberly & the Five Strange Goldfishes
The Enchanted Library
The Misadventures of Pirate Pete
From Wheel To Web: 40 Remarkable Inventions

www.ingramcontent.com/pod-product-compliance
Lightning Source LLC
Chambersburg PA
CBHW031135130726

47988CB00006B/2389